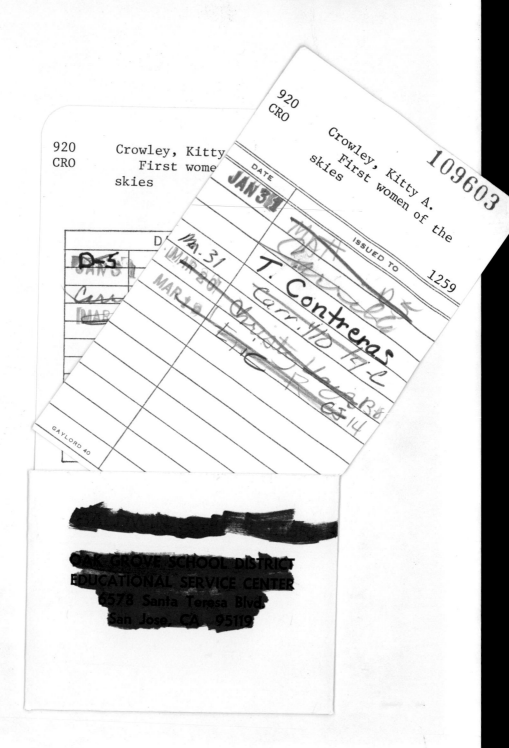

# First Women of the Skies

by
Kitty A. Crowley

Illustrated by
Russell Charpentier

**cpi**
contemporary perspectives, inc.

Copyright© 1978 by Contemporary Perspectives, Inc.
All rights reserved. No part of this book may be reproduced or utilized in any form or by any means, electronic or mechanical, or by any information storage and retrieval system, without permission in writing from the Publisher. Inquiries should be addressed to the PUBLISHER: Contemporary Perspectives, Inc., 223 East 48th Street, New York, New York 10017.

Library of Congress Number: 78-21907

Art and Photo Credits

**Cover illustration by Russell Charpentier**
Photos on pages 7, 39, and 44, courtesy of Emily Warner
Photos on pages 9, 13, and 21, The Bettman Archive, Inc.
Photo on page 23, NOVOSTI from SOVFOTO
Photos on pages 29, 31, 35, and 37, TASS from SOVFOTO
Photo on page 33, ARTKINO from SOVFOTO
Every effort has been made to trace the ownership of all copyrighted material in this book and to obtain permission for its use.

**Library of Congress Cataloging in Publication Data**

> Crowley, Kitty A. 1942-
>   First women of the skies.
>
>   SUMMARY: Brief biographies of five women pioneers of the air: Harriet Quimby, Mathilda Moisant, Bessie Coleman, Valentina Tereshkova, and Emily Warner.
>   1. Women air pilots — Juvenile literature. 2. Women astronauts — Juvenile literature. [1. Women air pilots. 2. Women astronauts. 3. Air pilots. 4. Astronauts.] I. Title.
> TL 539.C76      629.1′092′2    [B] [920]    78-21907
> ISBN 0-8945-063-2

Manufactured in the United States of America
ISBN 0-89547-063-2

# Contents

An Historic Flight     4

Chapter 1     8
   It All Started With . . .

Chapter 2     23
   First Woman in Space

Chapter 3     38
   A Pioneer of Today

A Final Word     48

# An Historic Flight

For most of the passengers this was just an ordinary airplane ride — Frontier Airlines Flight 12 from Denver to Saint Louis. It would take them to friends and family, or maybe to a business meeting. The crew knew that there was something just a little different about this flight, though. It wasn't anything to worry about. But it was something that made them all feel proud.

As the flight attendants closed the plane's heavy door they looked at each other. They nodded and even smiled a bit. Then one of the women knocked quietly on the door that led to the small, crowded cockpit. "Ready," she said softly.

A man came out and took the microphone a stewardess handed him. "May I have your attention, please?" he asked.

"Oh, no!" said a white-haired woman in the front of the plane. "He's going to tell us that something's wrong with the plane!"

The man with the microphone waited until the passengers were quiet. Then he spoke again. "Today," he said, "is a very special day for us. And this is a very special flight."

He smiled at the passengers. "Today Emily Warner is flying for us. She's the first woman to fly for a regularly scheduled American commercial airline. She'll be the flight engineer for your flight. We at Frontier are proud to have her flying with us. Thank you. And we hope you enjoy your flight."

The man put down the microphone and watched the passengers' faces. Some people looked surprised. Some whispered — not too quietly either.

"Well, *that's* something. It's about time," a woman said.

A man's voice was heard. "Why me?" he asked. "Couldn't she have picked some other flight for practice?"

Another man started to laugh. His loud voice could be heard all over the plane. "Well, I sure hope women pilots are better than women drivers. If they're not, we're all in trouble!"

Other people turned around, staring at the two men. Then one by one the passengers began to clap their hands. Soon all of them, even the two men, were clapping. They were welcoming Emily Warner aboard!

The rest of the trip was like any other flight from Denver to Saint Louis. The crew had known it would be. Takeoff and landing were perfectly smooth.

Emily Warner was finally in the cockpit of an airliner.

Nothing unusual happened in the air. But Emily Warner made history that day.

She really was the first woman to fly for an American airline. Emily Warner was not the first woman pilot, though. In fact there have been women pilots for almost as long as there have been airplanes.

# It All Started With . . .

In 1911 Harriet Quimby was 27 years old. She had finished 33 hours of flying lessons and had flown in the air for only 4½ hours. But on August 2, 1911, she became America's first woman pilot. In those days there were no official government licenses for pilots. Instead, certificates were given out by the Aero Club of America. Harriet Quimby's certificate was number 37, so she was one of America's first real pilots. And she was the first American woman of the skies.

Two weeks later another American woman was given a certificate by the Aero Club. Her name was

Harriet Quimby was the first American woman to get her pilot's certificate.

Mathilde Moisant. She and Harriet were close friends. In fact the women had taken lessons together at the flying school started by Mathilde's brother John. Soon Harriet and Mathilde would become famous as two of the bravest and most daring pilots of their time.

Flying was not for everyone in those days. The planes were light and not very strong. Most were made of only wood, canvas, and wire. It did not take much to damage a plane. Wind, ice, rain, or even the touch of a tree's branch — any of these could bring one of the small planes crashing to the ground.

Pilots had to work hard every moment in flight just to keep the planes in the air. Sometimes this was not easy. The pilots sat out in the open. Nothing protected them from the weather. With no safety belts, there also was the chance that pilots could fall out of the planes. Just as dangerous to pilots, many of the planes' engines threw hot oil in every direction. Only their goggles kept pilots from being blinded. And even then the oil became so thick on their goggles that pilots could hardly see at all.

No wonder so few people took to the skies in those days! But Harriet and Mathilde were like other early pilots. They loved flying. That it was dangerous and difficult never seemed to stop them. They even seemed to enjoy the thrill and danger of flight.

Like many other fliers, these two women wanted to set records. In the first days of flight everyone wanted to be the first to do something or go somewhere, the one to fly the farthest or the longest. Harriet and Mathilde were no different. On September 13, 1911, Mathilde set out on a cross-country flight. It was her 25th birthday. It was also the beginning of a trip that would set many records. During the flight Mathilde took her plane, *Lucky 13*, higher than any other woman had ever flown — 1,500 feet. Later she became the first woman to fly over Mexico City.

Harriet, though, was not about to let Mathilde set all the records. She had already been the first woman to fly at night — something even most men pilots did not do. Now Harriet wanted to set the record for the longest flying time by a woman. She quickly set that world's record.

Then a Frenchwoman announced that she would try to beat Harriet's record. Her attempt was to be made on a Sunday afternoon. Harriet's friends urged her to fly that day too, to try to keep her record. But Harriet had promised her father she would never fly on Sunday. Some things, he told her, were more important than flying. Harriet kept her promise. She wouldn't fly. The Frenchwoman set a new world mark that day.

Mathilde Moisant soon had her certificate and plane too.

A few months later Harriet was off on an even more dangerous flight. On April 16, 1912, she set out to fly across the English Channel. It was something no woman had ever done. Harriet wore her favorite necklace and bracelet for good luck. She also wore a pilot's outfit she had designed herself — a purple satin blouse with bloomers to match. High-laced shoes covered her feet. On her head was a purple satin hood.

Over her arm Harriet carried her goggles and a long leather coat. She would need these in the chilly fog over the Channel. She also carried a compass. It was the first time she would be using a compass in flight. But in thick fog, and with only water below, there would be no other way for her to know where she was going. The compass would be her only eyes during much of the trip.

As she took off Harriet watched the land slowly disappear behind her. Soon she was flying over the English Channel. In every direction, as far as she could see, there was only water. There were no landmarks to tell her where she was. Only the compass showed her where she was going. There wasn't even a way to call for help. If she crashed, no one would ever know. For now, though, everything was fine.

Suddenly Harriet was in trouble. The engine had stopped, leaving her plane without power. At any

moment the plane could fall and crash into the sea. Harriet worked at the controls, trying to restart the engine. The plane glided lower and lower, until Harriet could almost feel the icy water of the Channel. Then, just as suddenly as it had stopped, the engine started. Using the controls, Harriet slowly raised the plane over the water. She was safely in the air again.

It was cold over the Channel. Many times Harriet felt as if her fingers were too cold to work the plane's controls. But she flew on until she finally reached the coast. She had done it! She was the first woman to fly across the Channel.

That flight taught Harriet one thing, she believed. Flying was dangerous. But it could be done safely — if one was careful. Everything had to be carefully checked. Nothing could be left to chance. She even wrote a story about this for *Good Housekeeping* magazine. In it she said that flying could be a perfect sport for women. Women, like men, should just take care and check all their equipment.

There were some things, though, that couldn't be checked. In the early days of flying, chance was a part of every pilot's life. Luck had to be on the pilot's side at all times.

On July 1, 1912, Harriet set out for another record. This time she wanted to break the men's speed record.

Mr. Willard, the man who had organized her record attempt, flew with her. Harriet's plane climbed higher and higher, quickly building up speed. Soon she was nearing the record.

Then something happened. Even today no one knows exactly what went wrong. Mr. Willard suddenly fell from the plane. His fall threw the plane off balance. In a few moments Harriet too had fallen from her plane. The people below watched in shock as Harriet and her airplane fell 2,000 feet to the water below. She had been flying for just 11 months when she died.

Luck, though, seemed to be on Mathilde Moisant's side. Only a few weeks after Harriet Quimby's death Mathilde's plane crashed in Louisiana. She escaped with only a few bruises. A few days later she crashed again. This time it was in Texas, and once again she was lucky. A crowd had gathered, hoping to see the famous flyer. When she came in for her landing Mathilde found her way blocked by people. Her plane burst into flames as she tried to keep the plane from crashing into the crowd. But once again Mathilde escaped.

Mathilde wanted to keep flying. But her family pleaded with her to stop. Her brother John had already died in a plane crash. And Mathilde had had so

many narrow escapes. In 1912, at the age of 26 and after only a year of flying, Mathilde Moisant retired from the air.

Bessie Coleman, born January 26, 1893, wanted to learn to fly too. Right after World War I she tried to get someone to teach her. But she could not find a teacher in the United States. Bessie was black. She did not give up, though. She learned to speak French, went to Europe, and learned to fly from French and German pilots!

When she was 29 she started flying in air shows as a stunt flyer. She did figure eights and loops. She would fly the plane high into the air. Then, diving downward as fast as she could, she came so close to the ground that the spectators could see her face. Only at the last moment would she soar back up in the sky. Crowds were amazed at her daring stunts.

On April 30, 1926, Bessie flew for a show in Florida. The money raised would help the Negro Welfare League of the town. She had been in the air for ten minutes when the plane began to dive. At first the crowd thought it was another of Bessie's stunts. They expected things like this from Bessie Coleman. But then, as the plane flipped over, they realized the awful truth. Like Harriet Quimby, Bessie Coleman fell to her death. Bessie Coleman's great dream was to open a

When Bessie Coleman landed her plane people would rush to welcome her.

school where black people could learn to fly. She died before she was able to start it.

When people first began flying, planes were not very safe. Thanks to the first fliers much was learned during those early years. Slowly planes were made safer and easier to fly. As time passed flying became the safe, easy way to travel that it is today.

Harriet Quimby, Mathilde Moisant, and Bessie Coleman were three of these brave pioneers of the sky. The history they began still continues. There are still pioneers of the sky — and there probably always will be.

Valentina Tereshkova climbed aboard her spaceship — she would be the first woman in space.

# First Woman in Space

Valentina Tereshkova was taking the longest walk of her life. Ahead of her stood the Soviet spaceship *Vostok 6*. She walked toward it, wearing the orange spacesuit and white helmet of a Russian cosmonaut. It was the morning of June 16, 1963.

"My own spaceship," she thought. "I can hardly believe it." The moment when she would be lifted off into space was getting closer and closer. Her dream was about to come true.

Many years had passed since the flights of Harriet Quimby, Mathilde Moisant, and Bessie Coleman. But Valentina's thoughts were much like the ones those women had had. "If everything goes as planned," she said to herself, "I will be the first woman to pilot a ship into space."

As she rode in the elevator up to the cabin of *Vostok 6*, Valentina thought back over everything that had led up to this moment.

Valentina was born on a farm in Russia on March 6, 1937. Her father was killed in World War II when she was just three years old. Five years later her family moved to Yaroslavl, an old city on the Volga River.

The years passed quickly. Valentina's mother found work in a cotton mill, where she ran a loom. Valentina's life was filled with schoolwork and the games she played with her brother and sister. When she was ten she joined the Young Pioneers, a club for young people in the Soviet Union.

After finishing high school Valya (Valentina's nickname) got a job in a tire factory. She worked there four months, until she had a chance for a job at the cotton mill where her mother had worked. Valya was proud of that and used her first paycheck from the mill to buy her mother a present.

When she was 20 Valya began attending a technical college. She was still working at the cotton mill, so she could only go to school part-time. But what she learned at the college would lead to a better job at the mill. At the same time she joined the Young Communist League, one of the most important clubs for young people in the Soviet Union.

Valya quickly finished her courses at the college. She was even elected head of the league's branch at the mill. Valya, everyone said, had a bright future. It was not long before she was invited to join the Communist party. It was an important honor to her. It showed how much she was respected and trusted by others. It also meant that she was certain to have a successful career.

One morning, though, Valya had a real surprise for her mother. She got up very early. "Can't you sleep, Valya?" her mother asked. "Is something the matter?"

"Nothing's the matter, Mother. I was going to surprise you tonight. But I'll tell you the news now.

I'm going to learn to skydive. I joined the club at the mill."

"Oh, no, Valya," her mother said. "It's not safe. Besides, it's not a sport for girls."

"It's a sport for anyone who wants to do it. Oh, think how exciting it will feel to float through the air."

Now Valya's adventures really began. She mastered many difficult jumps. But one day, near the end of a jump, she looked down. Something was wrong. "Oh, no!" she said to herself. "I'm over the water. It's not supposed to be there! No, I mean, *I'm* not supposed to be *here*!"

A moment later Valya was in the water. "Help!" she cried. But there was no one to hear her. She struggled out of her parachute. Somehow she made it back to shore. Valentina was lucky that day.

Her family asked her why she kept on jumping. Valya thought a moment. Then she answered, "I love the excitement. You know I like to read about adventure. Well, I also like to have my own adventures."

Along with most of the world, Valya was excited to read about Yuri Gagarin, the first man to go into space. In 1961 this Soviet cosmonaut circled the earth in a spaceship called *Vostok 1*.

Unlike most people Valentina did more than read about it. She said to her friends, "This is for me. I want to be a cosmonaut more than anything else in the world — or out of it!"

Her friends said, "Valya, are you crazy? You're a woman. You've never even flown a plane. You could get killed!"

Nothing could change Valya's mind. She wrote a letter, asking if she could join the training program for cosmonauts. Then she waited for an answer. It seemed to take forever, but finally it came. She was admitted to the program even though she was not a pilot. Perhaps all that skydiving had paid off!

Valentina began her training in March 1962. She soon found out that training was like going through a house of horrors that never seemed to end. Space travel was new, so cosmonauts were trained to be ready for anything that might happen. The training was not for people who were weak or easily scared.

First Valentina had to sit in a chair. The chair was swung and spun around. Some people got sick to their stomachs — but not Valya. This was the first of many tests. Would she be able to pass the others? If she failed even one test she could not be a cosmonaut.

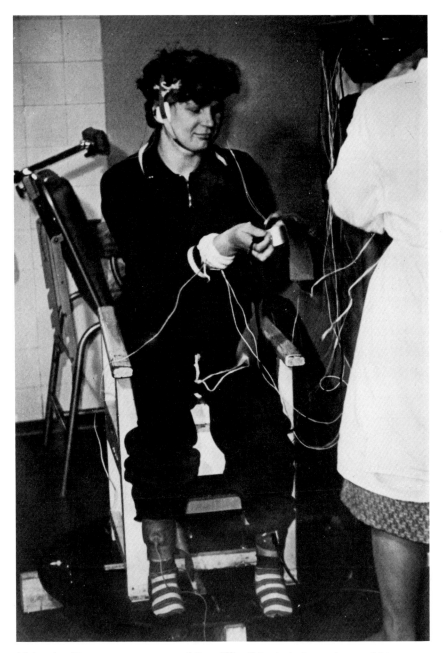
Valya had to pass every one of the difficult tests before she could become a cosmonaut.

Next there was a room that was hotter than anything she had ever felt. The outside of a spaceship got so hot when it returned to earth that it would begin to burn. This hot room was to prepare her for that. Again Valya passed the test.

An easy test for Valya was the skydiving. She had to land on the ground and then in the water. Valentina laughed. "Who would have thought that landing in the Volga River by mistake would help me become a cosmonaut?"

As part of her training she even had to learn to fly a plane. But there were two tests so awful that they had been given special names by the students. One was the "devil's merry-go-round." Valentina was strapped to a seat and spun until her body felt five times as heavy as it usually did. This is the way she would feel during takeoff and landing.

If that was not enough, there was the "chamber of horrors," a room that was totally dark and quiet. It was important to find out how long cosmonauts could stand to be in it, because space is black and without sound.

While she was in the "chamber of horrors" she thought about her idol, Yuri Gagarin. He had become one of her teachers. After working with her he said, "Valentina has amazing abilities .... She's not afraid to say, 'This is something I don't know about. Help me.'

The training and testing went on for months.

And we help her. She may be thin and weak looking at first glance, but she has great strength, energy, and willpower."

At last the long hours of training were over. Valentina had passed all the tests. But she was not the only woman to pass. There were others who wanted to go just as badly as Valentina. But in the end it was Valentina who was picked from among all the women to be the pilot-cosmonaut for *Vostok 6*. "Today," she thought, "I know what real excitement is!" She was proud that she had been chosen. She even liked the code name she would use during her flight. It was "Chaika," the Russian word for sea gull.

Soon that day in June 1963 came. Valya climbed into *Vostok 6*. Ground control was talking to Valentina as she sat in the spaceship. "Thirty minutes until takeoff." Soon she would be soaring through the sky like a sea gull. Then she would join a cosmonaut already in orbit. Two days earlier, Valeri Bykovsky had gone up in *Vostok 5*. His code name was "Hawk." The plan was for Sea Gull to fly in almost the same path as Hawk.

The countdown continued. Valentina was ready. Everything she had learned would be put to the test in just 60 seconds. This is why she had been working so long and so hard.

Valya felt the rocket's engines fire. *Vostok 6* slowly lifted off the ground. She was on her way. At first she felt as if she was back on the "devil's merry-go-round." Her body pressed back against the seat. But within five minutes she had left the atmosphere. Her body no

Valya sat in the spaceship, waiting for the countdown to begin.

longer felt heavy. In fact she felt as if she weighed nothing at all.

Her ship drew closer to *Vostok 5*. Sea Gull talked to Hawk. Together they sent a message to earth. "We are at close distance from each other. All systems in the ships are working excellently. Feeling well."

Valentina had never been so happy. She had made it! She was in outer space. Sea Gull was moving 300 times faster than the cars on a highway. But to Valya it seemed easy — and almost fun.

"Are you all right? Answer us, please." Ground control was trying to talk to Sea Gull. There was no answer. Ground control called Hawk. "Sea Gull does not answer. Please try to talk to her." Hawk tried too. Again there was no answer.

Suddenly Valya woke up. "Where am I?" she asked herself. "What is going on?" Then she remembered. "I'm sorry," she called to Hawk and ground control. "I know I wasn't supposed to fall asleep. I won't let it happen again — unless it's a time when I'm supposed to."

Work kept Valentina busy during her flight. She wrote in her logbook, describing what the earth looked like from space. She took pictures of the sun. And she worked the ship's controls. She also was on television, showing everyone what it was like to be weightless. Millions watched as she let her pencil and logbook float around the cabin.

She completed a trip around the earth once every 88 minutes. When she passed over the Volga River she thought of her mother. She called ground control. "This is Sea Gull. Please tell my mother not to worry."

"Sea Gull" showed people on earth what it was like to be weightless in space.

Later ground control thought they heard singing. It was Valentina. "I'm singing to Hawk," she joked. "I don't want him to get bored."

The plan had been for Valya to fly for one day. Then, if she felt all right, she could stay up longer. "I'm feeling fine," Sea Gull told ground control, "I'd like to stay up longer." She could stay up two more days, she was told. Then she would land at about the same time as Hawk.

Sea Gull came down first, at 11:20 A.M. She had flown in space for 71 hours and had gone around the earth 48 times. As she came back into the atmosphere, Valentina could see flames outside the spaceship. She was not afraid. She had known it was going to happen. When she was close to the ground she opened the door of her ship. She jumped, skydiving to safety. After she landed, someone asked her how she felt. "Only my nose is bruised," Sea Gull answered.

Today Valya no longer flies in space as a cosmonaut. But she still is an important person in the Soviet Union. Recently she became a doctor of aeronautical engineering. In 1974 Valentina Tereshkova was named a member of the Supreme Soviet Presidium. There she helps make laws for the whole country. She also is a member of the Central Committee of the Communist party, one of the most important jobs in the Soviet Union.

Valentina Tereshkova was the first woman to travel in space. She showed that women could work in the cold darkness of space just as well as men. Like Harriet Quimby, Mathilde Moisant, and Bessie Coleman, Valya was one of the first women of the skies.

At last the flight was over — Valya and *Vostok 6* were back on earth.

# A Pioneer of Today

Like Harriet Quimby, Mathilde Moisant, and Bessie Coleman, Emily Warner was an American pilot. Her flight for Frontier Airlines was an important first for women fliers. Harriet Quimby had earned her pilot's certificate long ago, in 1912. Since then many women had become pilots. And Valentina Tereshkova had even flown in space. But it was not until February 1973 that Emily Warner became the first woman to fly for an American passenger airline.

For a long time Emily Warner had wanted to work for an airline. First she thought she would become a

Emily and her twin sister, Eileen, grew up in Denver, Colorado.

stewardess. That's what many women did in the 1950s if they wanted to work for an airline. At some point, though, Emily said to herself, "I don't want to ride in planes. I want to fly them. And I want to get paid for doing it."

Emily was about 20 years old when she first thought of working for an airline. As she did with most things, she talked it over with her family. Her idea must have surprised them. "Dad," she said, "I want to take flying lessons. I have a job. I can pay for them with the money I earn."

Emily's family lived in Denver. They were a close, happy family. But they did not have much money. Emily knew it would not be easy to sell them on the idea of spending her money on flying lessons. She tried hard, though. Finally her father said she could try a few lessons — just to see if she liked it. Meanwhile she would have to keep up with the business courses she had been taking. And she would keep her job too.

In February 1958 Emily began her lessons. It took two hours just to get to the Clinton Aviation School. And the lessons cost a lot of money — $13 an hour. Emily earned only $38 a week working in a department store. But somehow she always found the money for her lessons.

By May she was flying fairly well. By June she had made her first solo flight. But her parents wanted her to give up her lessons. It was too expensive — and Emily knew it. Emily also knew how important flying was for her. She had flown alone! Now she wanted to get her private pilot's license. Once again she talked with her parents, pleading with them to let her go on.

"I want to get a job with one of the airlines in Denver," she told them. "I'll try to get a job as a secretary. That way I can work with people who love flying as much as I do."

Emily's mother and father understood. It was hopeless! There was nothing they could do but agree.

The only job she could find was as a receptionist. It was not as good as she had wanted, and it was only part time. Still, it was at the airfield. Emily took it. And soon she had earned her private license. Then she started working for her commercial license. When she got that, she knew, she could get *paid* for flying!

In the next few years Emily spent as much time at the aviation school as she could, flying whenever she could. She became a secretary at the school. Then a local radio station called the school. They wanted to broadcast highway traffic reports from the air. Emily often piloted the plane for the man who made the reports. She logged hours and hours of free flying time this way. It was a big step toward her commercial license.

It wasn't long before Emily got that license, just as she had gotten all the others. But who was going to pay her to fly? At first only the aviation school did. She delivered airplane parts to other airports. Then she began picking up new planes and flying them back to Denver.

It was not long before Emily earned her flight instructor rating. This meant she could teach others to fly. Now Emily flew all the time. She could share her love of flying with others. Soon her whole family had flown with her. Her brother Dennis even got his own private license.

In 1966 Emily became assistant manager of the flight school. She also became an examiner for the Federal Aviation Agency. Now she was giving flight tests to other people who wanted to fly. They could get their licenses only when she decided that they flew well enough.

Until then most of Emily's flying had been in single-engine planes. After she became manager of the flight school she began to get experience in planes with more than one engine. It was not long before she got

Emily and Eileen are both in uniform now. Emily flies for Frontier, and Eileen is an officer in the Air Force.

her rating for those planes too. By then Emily had another goal. Like the others, this one took people by surprise — she wanted to be a pilot for a commercial airline.

Emily wrote to the three airlines that landed at Stapleton International Airport in Denver. Continental Airlines didn't answer. United Airlines sent her a form letter.

Things were not much better at Frontier. But she did get to talk to the chief pilot there. "Get more experience," he told her. "Do more flying across the country. Fly more multiengined planes. Get your airline transport rating. That's like having a college degree in aviation." He stopped for a moment. "Someday," he said, "one of the airlines will hire a woman as a pilot. But I don't know when."

Emily did what he advised. And each year she sent in her application, showing all she had done. But it was a bad time for the airlines. Business was not good. Instead of hiring pilots, airlines were laying them off! And when times got better, those pilots would get the first chance at new jobs. It didn't seem like Emily would ever get her chance.

But other things were happening in the United States during those years. Women started asking why they shouldn't get the same kinds of jobs as men. This *made* the airlines think about hiring women as pilots. It also made Emily even more determined.

In 1972 Emily had something new to put on her application. She had helped train men to become pilots for United Airlines. If she was good enough to teach them, she said, she certainly must be good enough to be a pilot herself!

Frontier Airlines was hiring again. Emily applied once again at Frontier. And every few weeks she would ask to be put in the next training class for Frontier pilots. Everyone at Frontier was used to seeing Emily. But she still didn't get a job.

But one day a friend called her. "Mr. O'Neil, the vice-president of flight operations, will be in his office today. I hear they'll be taking on pilots for a January training class. Get over here at one-thirty and talk to him."

That was all Emily needed to hear. She went over to Frontier Airlines to see Mr. O'Neil. The next day the airline called. "We'd like you to come and talk with us about a job as a pilot." She had three separate interviews and a flying test. Still she had no idea what Frontier would decide.

Finally Frontier asked her to come over. "Emily," she was told, "the job is yours if you want it." They asked her to think about it overnight. That night she laughed when she remembered how the airline thought she might need time to make up her mind. After all, she had been asking for the job for almost six years!

The newspapers wrote about how Frontier had hired Emily. Then they wrote about Emily's first flight. The Smithsonian Air Museum even asked for her first uniform.

They made a statue of her wearing her Frontier Airlines uniform. From flight engineer Emily went on to become captain of a De Havilland Twin Otter. After that, despite the fact that many Frontier pilots still didn't like the idea of flying with a woman, she became a first officer of a Boeing jet airplane. Emily still is not satisfied, though. "I hope someday to become captain of a larger plane," she says.

# A Final Word

Like other women pilots, Emily Warner did not find it easy to make her dreams of flying come true. But just like the women who came before her, she was determined and she refused to give up.

There are many years and millions of miles of flight between Harriet Quimby and Emily Warner. But they are years and miles made possible by these — and all the other — women of the skies.